PYTHON PROGRAMMING

A Smart Approach for Absolute Beginners.

A Step-by-Step Guide with
08 Days Crash Course

STEVE M. MANSON

Copyright 2019 - All rights reserved.

All rights reserved. No part of this book may be copied, changed or transmitted in any shape or making use of any strategies, such as photocopying, recording, or different electronic or mechanical systems, without the previous made approval out of the distributor, beside through distinctive feature of short references embodied in critical overviews and positive other noncommercial usages permitted via copyright regulation.

Before You Start Reading This Book?

We are a team of programmers who love to code. Few months back when we searched through Amazon, we realized that the Python books written for beginners had the following issues:

- ❖ More than 50% of the books were just a compilation of information from the internet.
- ❖ The remaining books were either too lengthy or had accuracy issues.
- ❖ The lessons were quite disorganized..
- ❖ There were no concrete concepts and explanation of OOP
- ❖ There were no tips and techniques from the expert coders.
- ❖ There was no understanding of the complex topics of Python.
- ❖ In short, Python was made so difficult for anyone intending to learn that the journey from the beginner to the advanced level was almost impossible.

Therefore, we decided to jump in and help the students as much as we can with this guide book.

What To Expect In This Book (And What Not To)?

This book is designed to take the students from a basic level to an advanced level. The book is written keeping in mind that the student has no prior knowledge about Python (or programming) and that he wants to become an expert in Python in lesser time. Therefore, we assured that the student saves his time and money with maximized learning experience. In other words, we choose not to put any BS in the book.

We believe that the best way to learn python programming is by doing it. Therefore, this book is intentionally designed to help you not only learn python but also apply it in the real world. Also, the structure and format of the book will help you grab the basic concepts of python within the next few hours.

What this book offers?

- Examples with each topic to help you clarify the concepts.
- Definitions, examples and codes are thoroughly tested and proofread.
- Easy-to-remember concepts throughout the chapters
- The interactive examples with outputs.
- Visual illustrations to help you install Python

The distinctive features include:

- The book builds a solid foundation for other programming languages too.

- The book is made conversational. The writer speaks as if he talks to the readers directly. It is done so that you won't get sleepy or bored.
- The book makes python programming as easy as ABC.
- The book makes start programming like a pro.
- The book doesn't need you to have any prior knowledge of programming.
- The book explains potential of Python in the real world.

Table of Contents

Part 1: An Introduction To 'Python' 1

Day 1: Introduction To Python 2

    1.1 What Is Python? 2

    1.2 Why Learn Python? 3

    1.3 Whom Is This Book For? 3

    1.4 Versions Of Python: 2 Or 3? 4

    1.5 Downloading And Installing Python 4

    1.6 Writing Our First Python Program: 'Hello_World.Py' 10

Day 2: Variables And Data Operations 14

    2.1 Under The Hood Of 'Hello_World.Py' 14

    2.2 Using Variables 15

    2.3 The '=' Sign 17

    2.4 Naming Variables 17

    2.5 Basic Operators 18

Part 2: Basics Of Python 23

Day 3: Basic Data Types In Python 24

    3.1: What Is Data Types? 24

    3.2: Booleans 25

    3.3: Strings 26

    3.3.1: String Methods 27

3.: Numbers 35

    3.4.1: Integers 35

    3.4.2: Floats 35

3.5: Type Casting: What Is It? 36

3.6: Comments 39

    3.6.1: Why Are Comments Important? 39

    3.6.2: How To Write Comments? 39

Day 4: Advanced Data Types In Python 41

4.1: Lists 41

    4.1.1: Accessing Items In A List 42

    4.1.2: List Operations: Add, Remove, Edit And More 44

    4.1.3: Organizing List Items 49

    4.1.4: Extracting Parts From A List 51

4.2: Tuples 53

    4.2.1: Iterations, Conditionals And More In Tuples 55

    4.2.2: Making Changes In Tuples 56

4.3: Dictionaries 58

    4.3.1: Writing A Simple Dictionary 58

    4.3.2: Working With Dictionaries 59

Day 5: Decisions, Loops, Conditionals 65

5.1: Conditional Statements 65

    5.1.1 If Statements 65

  5.1.2 'If-Else' Statements 68

  5.1.3 'If-Elif-Else' Statements 69

 5.2: Loops 71

  5.2.1: 'For' Loop 71

  5.2.2: 'While' Loop 74

 5.3: Break And Continue 76

 5.4: Exceptions 77

  5.4.1: 'Try...Except' Block 78

Part 3: Diving Deeper Into Python 84

Day 6: Functions And Modules 85

 6.1: What Are Functions? 85

 6.2: Writing Functions 85

 6.3: Parameter Values 87

 6.4: Returning Values And The 'Return' Statement 90

 6.5: More On Function Arguments 92

 6.6: Scope Of Variables 96

 6.7: What Are Modules? 98

 6.8: Writing Modules 98

 6.9: Importing Modules 100

Day 7: File Management With Python 102

 7.1: Handling Files 102

 7.2: Opening Files In Python 103

7.3: File Operations: Read, Write, Append 106

7.4: A Brief Discussion On Binary Files 111

7.5: Closing Files 111

Day 8: Object-Oriented Programming 113

8.1: Introduction To Object-Oriented Programming 113

8.2: Classes And Objects 114

8.3: Writing Classes 115

8.4: An Explanation On Classes (Code Breakdown) 116

8.5: Making An Instance: Objects 117

8.6: Accessing Attributes And Methods 118

8.7: Inheritance 120

8.8: Child Classes: Writing One 121

8.9: Importing Classes 123

PART 1

AN INTRODUCTION TO 'PYTHON'

DAY 1

INTRODUCTION TO PYTHON

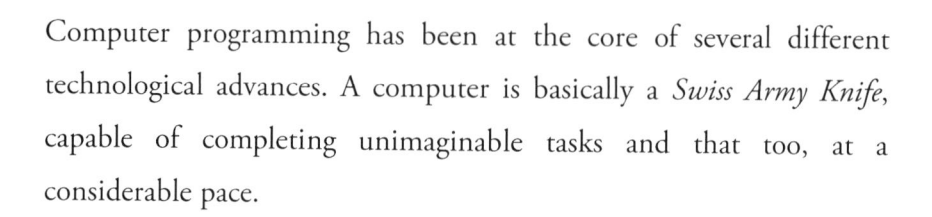

Computer programming has been at the core of several different technological advances. A computer is basically a *Swiss Army Knife*, capable of completing unimaginable tasks and that too, at a considerable pace.

In this modern era of computers and software, people do not need to click for hours and hours to repeat a task. Rather, a program, written in a language a computer understands, can quickly solve daily life problems.

But, how do you write these instructions for a computer? Through a programming language. This is where Python comes in.

1.1 What is Python?

Python is a programming language which helps us write instructions for computers and programs to follow. Whereas, the Python Interpreter is a software which takes our Python code and converts it into instructions that the computer can easily follow.

Python is a cross-platform programming language. This means it is supported by all major operating systems like Linux, Microsoft Windows, and MacOS. Luckily for you, the Python Interpreter is free to download. You can download it from https://www.python.org/downloads/.

1.2 Why Learn Python?

Python is a general-purpose programming language which was developed to solve a wide variety of problems. It is one of the most popular programming languages and loved by developers all over the world.

Its popularity also lies in the fact that its code is easy to write, read, and understand. The learning curve offered by Python is fairly simple (when compared to other languages) and tends to help beginners make a strong start in the field of Software programming.

Python is a fairly versatile programming language and can be applied to many use cases. From developing simple GUI (Graphical User Interface) applications to web applications, automation to scripting, data analysis to back-end development, Python can cover it all.

More recent use of Python has been in the domain of Artificial Intelligence and Data Science. Several different libraries, written in Python, assist developers in making Machine Learning models and solving computational problems.

1.3 Whom Is This Book For?

Within the course of this book, you will learn how to efficiently code programs and acquire the best practices of writing code, both in Python and other languages. Once you have completed this book, you will have a solid foundation of how programming languages work and learning a new language will be fairly easy for you!

1.4 Versions of Python: 2 or 3?

As of 2019, two versions of Python are available for use. These two versions include:

- ❖ Python 2 (2.7)

- ❖ Python 3 (3.7.3)

Although it's up to you to choose the version for your system, this book aims to help you get up to date with the latest version of Python (i.e. versions above 3.4.0)

Why you may ask? This is because the support for Python 2.7 is officially going to end at the end of 2019. Once the older version is deprecated, no updates will follow and there is no guarantee if code written in Python v3.7 will work on a system where Python v2.7 is installed.

1.5 Downloading and Installing Python

NOTE: The installation process states the version of Python at a few places. If you are viewing this at a later stage, rest assured, the process is going to be the same with a few minor differences.

Depending on your operating system, let's download the latest version of Python from the website: https://www.python.org/downloads/. As you may notice, Python is available for Windows, OS X, and Ubuntu for free.

Remember, the programs in the subsequent chapters are written in the latest version.

Now, if you have opened the website, there's an option stating "Download Python 3.7.3" (the version might differ if an updated version is released). Press the button, and scroll down on the new page which just loaded.

Microsoft Windows:

In the table of the "Files" section, you might come across Python installers for two types of computers; 32-bit and 64-bit. Now, if you are unsure what architecture your system follows, let's first figure that out.

On Windows, Select **Start** and open the **Control Panel**. You could either search for it or find another path to it. Once opened, click **System** and you can easily find your System Type from there. See if it's 32-bit or 64-bit and head back to your browser.

Now, find the appropriate file for your system and locate the **executable installer** for that version. For example, for users of the 64-bit system, the file is named **Windows x86-64 executable installer**. Click it, and the download should begin.

Once the download is complete, click on the installed file and follow the specified steps to install Python on your system:

1. Tick the **Install launcher for all users**
2. Tick the **Add Python 3.7 to PATH**
3. Press the **Install Now** button and let Python install. This should take a few minutes only.
   A. If you wish to turn off a few features or change installation settings, press the **Customize Installation** option. Although it is highly recommended to let the installer select the settings by itself for now.
4. Once installed, press the **Close** button, and you are done!

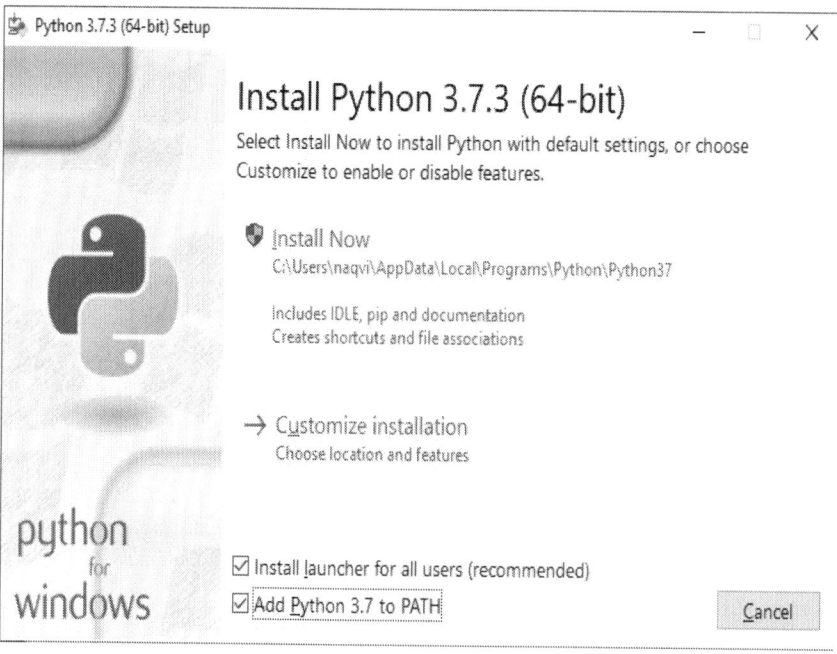

Figure 1: A typical installation screen of Python on Microsoft Windows

Now that we have installed Python, let's complete the process by setting up our programming environment on the terminal. Follow the following steps:

1. Open a command prompt window. Search for it in the **Start** menu or press **WIN + R**, and a **RUN** window will open. Type **cmd** and press **ENTER**.
2. Once the window opens, type **python** as a command in the window.
3. If you get a similar output (figure 2), your installed version of Python is detected and you just started a terminal session. Here, we will be writing our code!

Note: You can exit this terminal session by pressing **CTRL + Z**, or by simply shutting the window.

```
Python 3.7.3 (v3.7.3:ef4ec6ed12, Mar 25 2019, 22:22:05) [MSC v.1916 64 bit (AMD64)] on win32
Type "help", "copyright", "credits" or "license" for more information.
>>>
```

OS X:

If you are using Mac OS X, there's a chance Python is already installed on your system. You can verify this by starting a terminal session and typing **python** in it.

Go to **Applications > Utilities > Terminal**. A shortcut to opening the terminal is pressing **COMMAND + spacebar**, and then typing **terminal**. Press **ENTER**.

Now, once the window opens, type the word **python** in it. If the output displays the version of Python and a prompt starting with **>>>**, you have Python already installed on your system.

If there's no such output, begin by downloading the appropriate .dmg file for your system. Again, your system will either be a 32-bit or a 64-bit computer system and downloading the correct version is very important.

Find out your system type by opening the Apple menu. Then, select **About This Mac > More Info > System Report > Hardware.** Take a look at the fields which are displayed here. If the **Processor Name** field displays Intel Core Solo or Intel Core Duo, you have a 32-bit system. If not, you have a 64-bit system.

Once the file is downloaded, proceed with your installation by following these steps:

1. When the DMG file opens in a window, click the **Python.mpkg** file. If you are prompted for your password, enter it.
2. Click **Continue** to continue through the welcome screens.
3. **Accept** all licenses by pressing the button.
4. Select your hard drive (possibly HD Macintosh), and click **Install**.

Repeat the steps we did at the start for verifying for Python. If the command displays the version of Python and a prompt, you are good to go!

Ubuntu or Linux:

Apart from general-purpose uses, Linux systems are built from a developer's perspective and optimal for programming. In most cases, Python is usually installed by default.

Let's ensure this, by updating our system first then checking for the version of Python. Follow these steps:

1. Update and upgrade your system with the following commands: (type in the terminal)

   1. $ sudo apt update
   2. $ sudo apt -y upgrade

2. Once done, type the following command in your terminal to check the version of Python3 installed on your system.

   1. $ python3 -V

If your output is something similar to this:

Python 3.7.3

Python is already installed on your system and you are good to go! (the version might differ if your system is not updated).

NOTE: Throughout this book, **python** might be used to start the terminal or run our file. Remember, Linux has both versions of python installed. If you do not type **python3** as your command, Python v2.7 might try to run the file. It is recommended to type **python3** whenever you see the keyword **python** typed in the terminal.

3. If Python is not installed, type the following commands in the terminal and verify the installation again:

> 1. $ sudo apt-get install python3
> 2. $ sudo apt-get install idle3
> 3. $ sudo apt-get install python3-pip

1.6 Writing our first Python Program: 'hello_world.py'

By now, you must have successfully installed Python on your system. The Python Interpreter is the software which usually runs your programs in the back, whereas, the *Interactive Development Environment (IDLE)* is a software where which allows us to write our Python code in.

If you noticed, IDLE came with the installation of Python. Let's start the terminal up.

- ❖ On Windows, click the Start button, type **IDLE** in your search box or look for it in your programs. Double click or select **IDLE** (Python GUI).
- ❖ On Ubuntu, open your terminal and type **idle3** to run an instance of IDLE in your terminal.
- ❖ On Mac OS X, open your **Applications** list from the Finder's window. Click on **Python V3.X** (X is the specific version number), and then press the **IDLE** icon.

No matter which operating system you choose to use, IDLE is probably going to look and display the same things on startup.

Now that we have our environment set up and ready for service, let's begin by writing a simple program. In the world of programming, the "Hello World!" program is considered as the traditional beginning step.

In this program, we will be writing a line to display the text "Hello World!" on our screen. Now, I will be telling you two different ways of running the code we write in Python.

For our first method, head over to the IDLE window and type the following command:

1. print("Hello World!")

What just happened there? **"Hello World!"** is the output to your simple command. This shows that our installation was successful and our code was error-free.

Although this method of running our python code in the terminal results in correct outputs, it is not recommended for larger programs. In development, it is a good practice to keep your code stored in a file so you can run it as and when you want. The IDLE terminal doesn't keep a log of the commands and their outputs. It is best for running simple programs and testing if their outputs are correct.

For our second method, we will be needing a text editor. By default, each operating system has its own text editor. For example, on Windows, the **Notepad** application can be used to write text.

Although we are covering Windows, the steps still apply to all operating systems. Open your notepad and write the following command:

> 2. print("Hello World!")

Save your file with an appropriate name. I'll save it with the name **hello_world.py**. If you notice closely, with the file name, I appended the keywords "**.py**". The **.py** is the extension used to identify files written in Python.

By default, the Windows OS doesn't show the file name extensions. To turn it on, open your File Explorer, press the File option in the navigation bar on the top. Press the option "Change folder and search options". Open the View tab. In the advanced settings tab, scroll down to find the option **"Hide extensions for known file types"**. If it's ticked, untick it by clicking on the tick. Apply your changed settings and close the window.

Right click on your **"hello_world.txt" file** and rename it to **"hello_world.py"**. Let's run our code once again and see if it works.

Once saved, open your terminal. Change your directory in the terminal and head to where you saved your file. For example, if you are on the desktop, and the file is in a folder named **"PythonFiles"**. Here's the list of commands you have to type in your terminal:

> 1. cd PythonFiles
> 2. python hello_world.py

The "cd" command helps in changing directory.

In this method, we use the python interpreter to run our file. The output of this command should be the same, **"Hello World!"**.

Congratulations!

We just put our first step forward towards learning Python!

DAY 2

Variables And Data Operations

2.1 Under the Hood of 'hello_world.py'

We wrote our first line of Python in the last chapter. Now, let's take a deeper look at how Python understands our instructions, forwards it to the computer, and finally, displays our message.

As it so happens, Python does a fair amount of work in the back, most of it we are unaware of when it runs even a simple program like **hello_world.py**:

```
3.   print("Hello World!")
```

Now, when you run the file **hello_world.py** on your system, the file first goes through the Python interpreter, which breaks down the program, reads it word by word, and finally determines what each word should do.

For example, when the interpreter starts reading the first line, it comes across the word *print*. Now, some keywords are reserved in Python and have a specific meaning. The keyword *print* is a function (we'll go over functions later), which allows us to print (to the screen) whatever we put inside its parenthesis.

If you are working in an editor, you may have noticed that different parts of your code are highlighted differently. This feature of editors is called *syntax highlighting* and helps us understand specific parts of a Python program. For example, the ***print*** function must be a different color than our message "Hello World!", because one is a Python function and another is just a phrase. We'll learn more about this as we go.

2.2 Using Variables

Programs use more data than simple text messages or phrases. But, if you look back at our program, we simply ask it to return some phrase for us, and it is not stored anywhere.

This is where variables come in. Let's start by incorporating variables in our own program and explore it as we go.

Create a new file called **variables.py** and add the following lines to it:

```
4.   data = "Hello World!"
5.   print(data)
```

Now, run this program and let's see what does the interpreter output for us. If you typed it perfectly, the output should be:

Hello World!

So, what just happened here? We added a variable called **data** here and it has a value of **"Hello World!"**. Generally, each variable holds a value which signifies the information which the variable represents. In our case, it is our message to the world.

Let's analyze the program now. The Python interpreter first associated the phrase "Hello World!" with the variable **data**. On the second line, it prints the **value** held by the variable **data** but not the phrase **"data"** itself. Why? This is where the double quotation marks come in.

The quotation marks indicate the start and end of a phrase which have no relevance to Python and are text phrases only understood by us humans. Variables, on the other hand, store this phrase in them and can be used anywhere in the Python program.

Let's add on to our current program by modifying our file and adding two more lines to it:

```
6.  data = "Hello World!"
7.  print(data)
8.
9.  data = "This is just a phrase"
10. print(data)
```

Can you guess the value of data? Don't worry. Variables are called variables because their values can vary and there's no objection by Python if we change its value. So, when we change the value of our variable **data**, Python simply updates the current value of the variable and prints it when asked to. So, you should get the following output:

Hello World!

This is just a phrase

2.3 The '=' Sign

In mathematics, the "=" symbol is used to indicate equality. In programming, however, the "=" symbol is used to **assign values to a variable**. It is also called the **assignment operator** in most programming languages.

If you look back at our program, we *assigned* the phrase to our variable **data**. So, for the entirety of the program, the variable holds the value we assigned to it, unless, we change it ourselves (which we tried in the sample example in section 2.2).

2.4 Naming Variables

Make sure you name your variables rightly and use them properly. Here's the list of a few rules you should take care of:

- Variable names should not contain spaces in between. For example, the name "a variable" is not correct. Instead, you can connect these using underscores. This is the correct version: "a_variable".

- Only letters, numbers, and underscores are allowed in variable names. However, variable names may not start with numbers. Here are a few examples which are correct: "a_message", "messenger9", "no_3". These, however, are not correct: "9messages", "12people".

- We discussed keywords and how Python has specific meanings for them. So, do not use keywords or function names as variable names for your data.

- Variable names should be meaningful and describe the data held within them. For example, if you wish to convey a message "Hi!" to the screen. A good variable name may be "message", or "greetings".

2.5 Basic Operators

Operators are some symbols which help us in performing mathematical operations. If you consider a simple expression "5 + 11", here, the values 5 and 11 are **operands**, whereas, the "**+**" symbol is an operator.

Python supports many types of operators. Here are a few operators we will be using in our programs:

- **Arithmetic Operators**

Arithmetic operators are used to solve expressions and mathematical operations like addition, multiplication, etc.

Just like we assign phrases to our variables. We can also initialize them with numbers. Here's an example:

```
1. x = 15
2. y = 5
```

We will discuss numbers and the types of numbers supported by Python in great depth in the coming chapters.

Here are a few arithmetic operators with their examples:

Operator	Description	Example Code
+	Used to add two variables or operands	x + y
-	Used to subtract two variables. Subtracts from right to left.	x – y
*	Multiply two values or variables	x * y
/	Divide two operands. The left operand is divided from the right one	x / y
%	Modulus Operator – Used to calculate the remainder of the division of two operands. Left is divided from the right operand	x // y
**	Exponent Operator – Used to calculate powers. The left operand is raised to the power of right operand's value	x ** y

- **Comparison Operators**

Comparison operators are used to compare two values and decide the relation between them. They are also called **relational operators.**

Let's assume we have two variables with the following values:

1. x = 15
2. y = 35

The following table discusses a few comparison operators and their description:

Operator	Description	Example Code
==	Used to check for equality. If two operands have the same value, it returns True.	x == y (False)
!=	Used to check for inequality. If two operands have different values, it returns True.	x != y (True)
<>	If two operands have different values, the output is True.	x <> y (True)

	Similar to the != operator	
>	Greater than symbol. If the left operand is greater than the right operand, it returns True.	x > y (False)
<	Lesser than symbol. If the left operand is lesser than the right operand, it returns True	x < y (True)
>=	Greater than equal symbol. If the left operand's value is greater or equal to the right operand, it returns True	x >= y (False)
<=	Lesser than equal symbol. If the left operand's value is lesser or equal to the right operand, it returns True.	x <= y (True)

- **Logical Operators**

Assume the same variables and the values assigned to them. Here are a few logical operators supported by Python:

Operator	Description	Example Code
and	Logical AND operator. If both operands are true, the output is True.	x and b – returns True
or	Logical OR operator. If any one of the operands is true, the output is True.	x or b = returns True
not	Reverses the logical state of the operand. From True to False, and False to True.	not (a or b) – returns False. If 'not' is removed, the answer becomes True.

PART 2
BASICS OF PYTHON

DAY 3

Basic Data Types In Python

3.1: What is Data Types?

From our discussion in the last chapter, we know that variables are used to store the data we use in our Python programs. But, how do these variables actually work?

Each system has a Random-Access Memory (RAM), which is used to store temporary data. When we declare a variable as such:

```
1.  x = 10
```

What happens in the back is that Python reserves a small segment of our RAM for us and it is used to store that value.

But, how does the Python interpreter decide how much space to allocate for our variable? This allocation of space is dependent on the size of the variable, which in turn, is dependent on the *data type* of the variable.

To make our program more efficient and faster, we need to assign memory carefully and use the right data types for our variables.

3.2: Booleans

One of the simplest data types in Python and other programming languages is the bool or the Boolean data type.

A Boolean variable can only have two possible values; True or False. Such variables are used commonly to test for conditions.

For example, if you wish to know the state of the web server on in your Python program, you could write something like:

1. webServerOpen = True
2. lockdownState = False
3. underMaintenance = False

If you are unsure about the data type of a variable, Python allows you to easily access the data type using the **type()** function.

If you run the following code next,

1. print(type(webServerOpen))
2. print(type(lockdownState))

You will get the following output:

```
>>> type(webServerOpen)
<class 'bool'>
>>> type(lockdownState)
<class 'bool'>
>>>
```

The function returns the data type of the variable which is sent inside its parenthesis. The values inside the parenthesis are called *arguments*. Here, 'bool' represents a Boolean variable.

Boolean expressions will be further discussed in the next chapter when we explore conditionals. Practical examples will definitely clear help in understanding such topics.

3.3: Strings

Remember, in our first program, when we printed "Hello World!". We called it a phrase or text which is not built into Python but can be used to convey messages from the user. This text or similar phrases are called **Strings** in programming languages.

Strings are nothing more than a series of characters. Whenever you see something enclosed with double quotation marks or single quotation marks, the text inside is considered a string.

For example, these two variables that I've just declared are both Strings and perfectly valid.

1. a = "This is an example of a string and I'm using double quotation marks"
2. b = 'This is also an example of strings and now, I am using single quotation marks'

What does this flexibility offer? If you take a look at the variable 'a'. I've used a contraction and a single quotation mark inside my string (which uses a double quotation mark).

Here are a few examples of quotes and sentences where I've used contractions:

1. string1 = "I'm an amazing person"
2. string2 = 'I told him, "Work hard!"'

3.3.1: String Methods

Methods are simple actions that we can ask the Python interpreter to take for us. Methods require variables or a piece of data on which they can work and produce the desired result. Let's first write some code and we'll explore it as we go.

Changing the Case of Strings:

Create a new file called stringMethods.py (or alternatively use IDLE) and run the following code:

1. username = 'john doe'
2. print(username.title())

If you run the code, you should get this output:

John Doe

In the first line, we declare our variable and assign a string to it. In the second line, we start by printing something. In the parenthesis of the print function, we use the method **title()** on our variable.

Methods let us perform an action on our data. In this case, the title method displays each word of our variable in capital letters. Since the method acts on our variable, we connect the variable to the method using a period or '.' symbol.

Methods are identified by the set of parentheses and take in additional information to act on the variables they are connected to. Since the title method doesn't need any other information, the parenthesis is empty.

To deal with character casing, several other methods are available. Let's test the **upper() and lower() method** by adding the following code to your file. Then, run it:

1. username2 = 'John Doe'
2.
3. print(username.upper())
4. print(username.lower())

The output of these methods should be:

JOHN DOE

john doe

So, the upper method capitalizes each word, whereas, the lower method converts all characters to lowercase letters.

Concatenating and Combining Strings:

At times your program might require users to input their first name, middle name, and then the last name. But, when you display it back, these names are combined together.

So, how do we concatenate strings? It's simple. The "+" symbol is called the concatenation operator and is used to combine strings together.

Take a look at the following code:

```
1. firstName = "John"
2. middleName = "Adam"
3. lastName = "Doe"
4. fullName = firstName + middleName + lastName
5. print(fullName)
```

What's the output? **John Adam Doe**

If you want to run some other methods on the output, feel free. Here's a program which runs some tests on the variables:

```
1. firstName = "John"
2. middleName = "Adam"
3. lastName = "Doe"
4. fullName = firstName + middleName + lastName
5. print(fullName.lower())
6.
7. message = "The manager, " + fullName.title() + ", is a good person."
```

What would be the output? For the first print statement:

john adam doe – All lowercase letters. For the second print statement:

The manager, John Adam Doe, is a good person. – Concatenated a string on which a method was applied.

So, you can use these methods on every string in your program and it will output just fine.

Adding Whitespaces:

Whitespaces refer to the characters which are used to produce spacing in between words, characters, or sentences using tabs, spaces, and line breaks. It's better to use properly use whitespaces so the outputs are readable for users.

To produce whitespaces using tabs, use this combination of characters '\t'. Here's an example to show you the difference with and without the tab spacing.

1. print("This is without a tab spacing!")
2. print("\t This is with a tab spacing!")

Notice the difference in the output? Here's the output of the two lines above.

This is without a tab spacing

This is with a tab spacing

To add new text to new lines, you can add a line break using the combination '\n'. Here's an example to show you the difference a line break adds:

1. print('I love the following weathers: SummerWinter')
2. print('I love the following weathers:\nSummer\nWinter')

Once again, which output is more readable? Take a look at the outputs.

1. **I love the following weathers: SummerWinter**

2. **I love the following weathers:**

Summer

Winter

You can use both tabs, spaces, and newlines together and it would work just fine. Here's an example program to cater to both new lines and tabs being used together.

print("I love these weathers:\n\tSummer\n\tWinter")

Here's the output of the command:

I love these weathers:

    Summer

    Winter

Removing or Stripping the Whitespace:

To you, the phrases 'code' and 'code ' look the same right? To the computer, however, they are two very different strings. It treats this extra space as the part of the string; until, told otherwise.

Where does this produce a problem? When you are about to compare two strings to check if 'code' and 'code   ' are equal, you will think it's True, but the interpreter will always return a False.

If you think your string has extra space to the right side of the string, use the **rstrip()** method. Let's take a look at an example. I'll declare a variable, show you the whitespace, and then remove it to see the difference. Here's the code:

```
1. favoriteWeather = "Summer    "
2.
3. print(favoriteWeather + "is lovely")
4.
5. favoriteWeather.rstrip()
6.
7. print(favoriteWeather + "is lovely")
```

If you run these statements, you might expect the following output:

Summer   is lovely

Summer is lovely

Did you expect this output? Here's the actual output:

Summer   is lovely

Summer   is lovely

This is because the effect of **rstrip()** is temporary. Sure, it does cause the removal of whitespaces, but, to ensure it is permanent. You need to reassign it to the original variable. Here's the same example, continued to explain this concept:

1. favoriteWeather = "Summer   "
2.
3. print(favoriteWeather + " is lovely")
4.
5. favoriteWeather = favoriteWeather.rstrip()
6.
7. print(favoriteWeather + " is lovely")

In line 3, rather than using the method only, I assigned it to the original variable which causes the permanent removal of whitespaces. Here's the new output:

Summer   is lovely

Summer is lovely – After the removal of spaces

If you think your string has extra space to the left side of the string, use the **lstrip()** method. Here's a small program to test the lstrip() method:

1. weather = "   cold"
2.
3. print("It is " + weather)
4.
5. weather = weather.lstrip()
6.
7. print("It is " + weather)

Here's the output of the lines:

It is   cold

It is cold

If you wish to remove whitespaces from both ends, use the **strip()** method. Here's an example to show you the removal of whitespaces from both ends:

1. message = "   It's so cold in London   "
2.
3. print(message)
4.
5. message = message.strip()
6.
7. print(message)

Here's the output of these statements:

It's so cold in London   (end)

It's so cold in London (end)

If you want to temporarily notice the effects of strip, you can use the method in a print function and see the effects. These strip methods are extremely useful in real-life applications when user data has to be cleaned and manipulated.

3.: Numbers

Numbers are an integral part of our daily lives and computers make use of the two binary digits; 0 and 1, a lot. Python has several different ways to treat numbers. Here are a few common types we'll be looking at:

3.4.1: Integers

Integers are numbers that are written without fractional parts. In Python, these numbers have the type **int** (just like bool for Booleans).

Here are a few integers and their declaration in Python:

```
1.  a = 2424
2.  b = 10101
3.  c = 9040
```

3.4.2: Floats

If you declare a number with a decimal point, Python will automatically consider it a floating-point number or a float.

All operations you could perform on integers, they can also be performed on floating point numbers. If you run the type function on floats, you'll get a type **'float'**.

Here's a program to show some operations on floats:

```
1.  w = 1.2 * 0.3
2.  print(w)
```

```
3.
4.  x = 10.4 + 1.6
5.  print(x)
6.
7.  y = 10.555 + 22.224
8.  print(y)
9.
10. z = 0.004 + 0.006
11. print(z)
```

Here is the output to all four print statements:

0.36

12.0

32.778999999999996

0.01

3.5: Type Casting: What Is It?

Another fact about Python is that it is a dynamically-typed language.

A weakly-typed or dynamically-typed language doesn't associate a data type with your variable at the time you are typing your code. Rather, the type is associated with the variable at run-time Not clear? Let's see an example.

x = 10

x = "I am just a phrase"

x = 10.444

x = True

When you run this code, it'll perform its desired actions correctly. Let's see what happens with the variable 'x' though. We've written four statements and assigned different values to 'x'.

On run-time (when you interpret or run your program), on line 1, 'x' is an integer. On line 2, it's a string. On line 3, it's a float, and finally, it's a Boolean value.

However, through typecasting, we can manually change the types of each variable. The functions we'll be using for that purpose are **str()**, **int(), and float()**.

Let's expand the same example:

1. x = 10
2. x = float(x)
3. print(type(x))
4.
5.
6. x = "I am just a phrase"
7. print("x: " + x)
8. print(type(x))

```
9.
10.
11. x = 10.444
12. x = int(x)
13. print(type(x))
14.
15. x = False
16. x = int(x)
17. print(x)
```

In this program, we've used everything covered in the last few lessons. All data types and converted them using our newly learned function.

In the first case, x is converted into a float and the type function does verify that for us. Secondly, the string is still a string since it can't be converted into numbers, int or float. Thirdly, we convert the float into an integer.

As an added exercise, if you print the newly changed value of the third case, you'll see that the value of x is: **10.** This is because the type is now changed and the values after the decimal point are discarded.

In the fourth case, we print the value of x which is False. Then, we change its value to an integer. Here, something else comes up. The output? **0.**

It's because, in Python, the value of True is usually any non-zero number (typically 1) and for False, it's 0. So, their integer conversions yield 1 and 0 for True and False respectively.

3.6: Comments

Comments are text phrases that are put in the code to make it understandable and readable for other coders, readers, and programmers.

3.6.1: Why Are Comments Important?

Comments are very important, especially when you're working with other programmers and they'll be reviewing your code sooner or later. Through comments, you can write a small description of the code and tell them what it does.

Also, if you have other details or personal messages which are relevant to the code, you can put them there, since the interpreter doesn't catch them.

3.6.2: How to Write Comments?

In python, there are two ways to write comments, and we'll be exploring both of them. For our first method, you can use the "#" symbol in front of the line you wish to comment. Here, take a look at this code:

1. # This line is a comment
2. # Count is a simple variable

```
3.
4.  count = 15
5.  print(count)
```

If you run this code, the output will be **15**. This is because the comments lines (starting with #) are not run at all.

Now, this method is fine if your commented lines are less i.e. do not span over multiple lines. But, if they do, hashing all of them is a waste of time. For our second comment, we'll enclose our commented lines is **three single quotation marks (''') and close it with three quotation marks as well.** Here's an example:

```
1.  '''
2.    This comment spans
3.    on multiple lines.
4.  '''
5.
6.  count = 15
7.  print(count)
```

Notice, we have to close our multi-line comment, unlike the single line comment.

DAY 4

ADVANCED DATA TYPES IN PYTHON

4.1: Lists

In cases where we need to store large data, we need larger data structures to store our data. The first collection (store of data), we'll be discussing is the List.

A list is a collection (container of data), which is heterogeneous in nature i.e. it can store all data types in a sequenced order. Here's a list to get you started:

```
1.  aList = [1, 2, "A string", 11.50, 'c']
```

Let's examine the syntax. Just like our variable, we need to write a valid list name (similar to variable names, should be meaningful), and then equate it to a list. The list collection starts and ends with **[and]** symbol respectively.

We see that our list is populated with values of all data types and if we iterate over it using a for loop, here's the code and output:

```
1.  aList = [1, 2, 3, "A string", 10.5, 11.50, 'c']
2.
```

```
3. for element in aList:
4.     print(element)
```

```
1
2
3
A string
10.5
11.5
c
```

Lists are mutable. After their declaration, we can safely make changes to them and all those changes will be termed valid by Python.

4.1.1: Accessing Items in a List

If you print your list using this command:

```
1. aList = [1, "String", 10.5]
2. print(aList)
```

```
[1, 'String', 10.5]
```

Now, to convert this list into a user-readable form, we need to learn how to access elements of a list easily. Since each element in a list gets a fixed position (we call it *index*), we can use indexes to access an element of a list.

To access list elements, write the index (position) of the element you want to access, and enclose it in square brackets. Here's how you can access the first element of our list:

```
1.  aList = [1, "String", 10.5]
2.  print(aList[1])
```

Try printing this. Here's the output:

```
String
```

If you've understood the syntax. Here's the next logical question which you might have. Why did Python print 'String' for me, when I asked it for the first element, which was '1'?

Python starts its indexing from 0 and not 1. So, when you access the first (1st) element, it gives you the second element. If you access the 0th element, here's where you'll get '1' as output. Here's the code and its output:

```
1.  aList = [1, "String", 10.5]
2.  print(aList[0])
```

```
1
```

Likewise, the indexing continues, 0,1,2.... so on. Sometimes, you might want the last element of the list. If you ask Python to print the '-1' element, it will return you the last element of the list. Here's an example:

1. aList = [1, "String", 10.5, "Drums"]
2. print(aList[-1])

```
Drums
```

Similarly, if you wish to access the second last element what would you do?

Simply decrement the value, -2. You can keep decrementing until you're at the first element.

1. aList = [1, "String", 10.5, "Drums"]
2. print(aList[-2])

```
10.5
```

4.1.2: List Operations: Add, Remove, Edit and More

Lists are mutable, as we discussed earlier. So, during run-time or before execution, if you wish to change the elements of your list, you can safely do so using a few methods.

If you have a list:

1. aList = [1, 2, 3, 4, 5]

And, you wish to add two more numbers to it (in the end), you can do so using the **append()** method. Here's the method in play:

```
1. aList.append(6)
2. aList.append(7)
3. print(aList)
```

Here's the output list which contains our newly added numbers:

```
[1, 2, 3, 4, 5, 6, 7]
```

However, the append method only adds it to the end. What if you want to push something to your list at the 3rd index?

You can achieve your goal using the **insert()** method. This method now takes two arguments, the index, and the element you want to add inside.

Here's the same example and we add the new numbers to different indexes:

```
1. aList = [1, 2, 3, 4, 5]
2. aList.insert(2,6)
3. aList.insert(5,9)
4.
5. print(aList)
```

In this example, we use the insert method on our list to add new values. As the first argument to our method, we provide the **index or the**

position where the new element is to be added. When we write 2, Python starts counting the elements from the start and adds on (starting from **1 not 0**). When 2 elements are encountered, it adds our new element right after it. So, you see 6 after the 2nd element. Similarly, you can add 9 at the 5th position. However, this 5th position is now from the new list (after adding 6), not the older one.

Here's the output:

```
[1, 2, 6, 3, 4, 9, 5]
```

Now, just like you added elements, you might want to remove them from your list and it is done using the **remove()** method. You can simply pass to it, the value or the element you wish to remove from your list.

Here's the method in action (keep adding the code in the same file if you will):

1. aList.remove(9)
2. #Here, I'm asking it to remove my newly added element, 9

Now, the output is a list without the element 9.

```
[1, 2, 6, 3, 4, 5]
```

What was to happen if my list had multiple elements which matched the element in the remove method? Only the first occurrence is removed.

Here's the output when I add three different 9's to my list:

1. aList.insert(2,6)
2. aList.insert(5,9)
3. aList.insert(1,9)
4. aList.insert(3,9)
5.
6. aList.remove(9)

```
[1, 2, 9, 6, 3, 4, 9, 5]
```

Again, this removes elements that match our given argument. What if I wish to remove it from a specific index? You may use the **pop()** method. The argument part of the pop method is optional. If you pass no index, it will remove the last element from your list. And, if an index is provided, that specific index element is removed.

Here's the example code to test pop using both methods:

1. aList = [1, 2, 3, 4, 5]
2. aList.append(9) # [1, 2, 3, 4, 5, 9]
3. aList.insert(3,9) # [1, 2, 9, 3, 4, 5, 9]
4. print(aList)

```
5.
6. aList.pop() # [1, 2, 9, 3, 4, 5]
7. aList.pop(3) # [ 1, 2, 3, 4, 5]
```

Now, you just saw, the pop method removed both from the end (no argument) and from the third position (given index). Here's the output:

```
[1, 2, 3, 4, 5]
```

To remove an item, you also have the **del** keyword which can be used to remove an element using its index. Here's how you can remove the 1st and the last element from the list we've just formed above:

```
1. del aList[0]
2. del aList[-1]
```

```
[2, 3, 4]
```

Want to empty the list altogether? Use the **clear** method and you're good to go. Here's the code when we run clear on the list from the last output.

aList.clear()

```
[]
```

Here's the output, an empty list.

4.1.3: Organizing List Items

Let's populate our list again using the following code, but this time, let's use strings instead of numbers:

states = ["MS", "TX", "MO", "MT", "NV"]

Now, at times, to ease our processing, we want to sort our data or organize it in a manner that makes it more readable and easy to use. For this purpose, we can use the **sort()** method to sort our list alphabetically.

Here's how you do it:

states.sort()

print(states)

By default, the sort method follows ascending orders. So, it sorts the list from A to Z (alphabetical order).

Here's the before and after output:

```
['MS', 'TX', 'MO', 'MT', 'NV']



['MO', 'MS', 'MT', 'NV', 'TX']
```

However, you can modify the sort method and sort your list according to your needs.

Wish to sort the list in descending order? (From Z-A). Pass the **reverse** parameter and set its value to True.

Here's how a list is sorted in reverse order:

1. states.sort(reverse=True)

```
['MS', 'TX', 'MO', 'MT', 'NV']



['TX', 'NV', 'MT', 'MS', 'MO']
```

If you wish to reverse the list without sorting it, you can use the **reverse** method.

1. states.reverse()

```
['MS', 'TX', 'MO', 'MT', 'NV']



['NV', 'MT', 'MO', 'TX', 'MS']
```

4.1.4: Extracting Parts from A List

If you're not interested in accessing a single element or the entire list itself, you can extract a small part of your list using an easy technique. It is called *slicing* and makes a separate group of elements from a list.

To slice a list, you have to specify the starting index and the ending index which you wish to work with.

Slicing, however, starts at the 0th index and ends at the last index - 1. So, keep this in mind. Here's an example covering a few types of slicing:

```
1.  teams = ['Warriors', 'Raptors', 'Cavaliers', 'Bucks']
2.  print()
3.
4.  # Print elements from 1-2 (3-1)
5.  print(teams[1:3])
6.
7.  # Print all elements from 1 to end
8.  print(teams[1:])
9.
10. # Print all elements
```

```
11. print(teams[:])
12.
13. # Print last 2 elements
14. print(teams[-2:])
15.
16. # Print first two elements
17. print(teams[:2])
```

Here's the output:

```
['Raptors', 'Cavaliers']
['Raptors', 'Cavaliers', 'Bucks']
['Warriors', 'Raptors', 'Cavaliers', 'Bucks']
['Cavaliers', 'Bucks']
['Warriors', 'Raptors']
```

If you specify the first index, Python starts from there and takes it till the end of the list. If you specify both indexes, Python starts and ends at that index - 1. If you use a negative index as your first index, Python will start slicing from the end. If no indexes are provided in between the range (:), all elements will be printed. This is also equivalent to copying your list.

Also, you can loop over the sliced list. Either store it in another variable (which also becomes a list) or simply use it in a for loop.

Here's how:

```
1.  teams = ['Warriors', 'Raptors', 'Cavaliers', 'Bucks']
2.
3.  # Store elements from 1-2 (3-1)
4.  newTeamsList = teams[1:3]
5.
6.  # Variable which has our sliced list
7.  for team in newTeamsList:
8.      print("Title favorites: " + team)
9.
10. # Run-time slicing
11. for team in teams[2:4]:
12.     print("These are also the title favorites: " + team)
```

Here's the output:

```
Title favorites: Raptors
Title favorites: Cavaliers

These are also the title favorites: Cavaliers
These are also the title favorites: Bucks
```

4.2: Tuples

A tuple is another collection that is used to store different types of data. Although a tuple is ordered, it is not mutable, unlike our list which we can modify anytime.

Tuples are made to resist changes and don't allow edits after declaration. A tuple can be created using two parentheses, like (). The values or data inside these two brackets will be the elements of the tuple.

Here's how you can make a tuple:

```
1.  aTuple = ('A', 'B', 'C')
2.
3.  print(aTuple)
4.
```

Here's the output to when we print our Tuple:

```
('A', 'B', 'C')
```

You can access the elements just as you would access the elements of a list. Here's an example:

```
1.  print(aTuple[0])
2.  print(aTuple[-1])
```

```
A

C
```

4.2.1: Iterations, Conditionals and More in Tuples

Although its items are immutable, you can still loop over a tuple using a simple for loop. Here's an example to iterate over the elements of our tuple:

```
1. for el in aTuple:
2.     print(el)
```

```
A
B
C
```

You can also use the in membership operator just like other collections.

Here's how we check if our Tuple contains a string 'Vain':

```
1. if "Vain" in aTuple:
2.     print("FOUND VAIN!")
3. else:
4.     print("No sign of 'Vain'")
```

```
No sign of 'Vain'
```

If you want to find out the length of your list or Tuple, you can do so using the **len() function.**

Here's an example to run the len function on our newly created tuple:

```
1. print(len(aTuple))
```

3

4.2.2: Making changes in Tuples

We discussed the immutability feature in Tuples but didn't really test it. Let's try to make a tuple and make changes to it.

Since we know it might throw an error, let's put it in a try-except block we covered in the last chapter. Here's the code and the output:

```
1.  languages = ('C', 'Java', 'C++', 'Python', 'Rust', 'PHP')
2.
3.  print(languages)
4.
5.  # Try to change the third element
6.  # Should return an error.
7.  # Let's try it in a try-except block
8.  try:
9.      languages[2] = 'Not C++'
10. except TypeError:
11.     print("Tuples are immutable. No add, delete, or edit!")
```

```
('C', 'Java', 'C++', 'Python', 'Rust', 'PHP')
Tuples are immutable. No add, delete, or edit!
```

Similarly, if you try to add or delete new items. You'll get the same error. Sure, you can use the **del** keyword to remove the entire tuple at once.

So, how can you make changes to a Tuple? Rewrite the data in it or redefine the tuple as a new set of values.

```
1.  languages = ('C', 'Java', 'C++', 'Python', 'Rust', 'PHP')
2.
3.  print(languages)
4.
5.  try:
6.      languages[2] = 'Not C++'
7.  except TypeError:
8.      print("Tuples are immutable. No add, delete, or edit!")
9.
10. languages = ('C', 'Java', 'JavaScript', 'Python', 'Rust', 'PHP')
11.
12. print(languages)
```

```
('C', 'Java', 'C++', 'Python', 'Rust', 'PHP')

Tuples are immutable. No add, delete, or edit!

('C', 'Java', 'JavaScript', 'Python', 'Rust', 'PHP')
```

4.3: Dictionaries

Dictionaries are yet another collection type in Python which can be used to store data in a key-value pair manner. They are similar to real-world dictionaries. Each key-value pair in a dictionary is separated by a colon (:), and each key is separated by a comma.

Dictionaries are data types that are mutable, indexed, and unordered. Unlike the list, the 0th index of a dictionary can be anything. So, it is better to ask for elements using their keys. Let's take a look at an example dictionary next.

4.3.1: Writing a Simple Dictionary

Dictionaries are written with curly brackets unlike the list and tuples with [] and () respectively. Here's an example dictionary to hold the count of wins:

1. winCount = {
2.   'John': 35,
3.   'Adam': 44,
4.   'Michaela': 23

```
5. }
```

If you print this dictionary, the output is:

```
{'John': 35, 'Adam': 44, 'Michaela': 23}
```

Let's take a look at example operation on dictionaries next.

4.3.2: Working with Dictionaries

Accessing the elements of a dictionary is possible using a few ways. If you know the key of the element, you can access it using two square brackets and put the key in between Here's how:

```
1. johnCount = winCount['John']
2. print(johnCount)
```

```
35
```

You may also get the same result if you use the method **get()** on your dictionary.

```
1. johnCount = winCount.get('John')
2. print(johnCount)
```

Unlike Tuples, Dictionaries allow you to change the elements without an error. You can simply reassign a new value to a key-value pair in a dictionary using the following code:

```
1.  # John won 5 more games
2.  winCount['John'] += 5
3.  print(winCount['John'])
```

Now, if you are unaware of the keys of your dictionary, you can loop through it to access the elements using the **for** loop (Iterates over elements one by one. We'll study this in detail again).

Here's how:

```
1.  for element in winCount:
2.      print(element)
```

```
John
Adam
Michaela
```

But, these are only the keys of your dictionary. You can access the values of your dictionary using the method we studied earlier:

1. for element in winCount:
2. print(winCount[element])

```
40
44
23
```

If this isn't appropriate, you can use the **values()** method in your loop to loop over the values only. Here's how:

1. for element in winCount.values():
2. print(element)

```
40
44
23
```

Since dictionaries are a key-value pair, you can also access them as key-value using the **items() method.** Here's how:

1. for key, value in winCount.items():
2. print(key, value)

```
John 40
Adam 44
Michaela 23
```

If you wish to check if a certain key belongs to a dictionary or is valid for use, you can use the **in** membership operator with a conditional statement to do that. Here's how:

```
1. if "John" in winCount:
2.     print("John is here!")
3.
4. if "Mike" in winCount:
5.     print("Mike is here too!")
```

```
John is here!
```

Since mike isn't available as a key, it returns False and nothing is printed. Like the tuple and lists, you can use the **len() function** to find out the length of your dictionary.

```
1. print(len(winCount))
```

If you want to add new elements to your dictionary, you can do that using the same square brackets and write the name of your key inside. Then, assign to it a value.

```
1. winCount["Mike"] = 4
```

> 2. winCount["Amy"] = 9

Try printing this dictionary and trying out the changes yourself.

Just like the list, you can use the pop method to remove an element from your dictionary. But, it will require the key you wish to remove, otherwise, an error will be thrown.

> 1. winCount.pop("John")

You can remove the last inserted item using the **popitem()** method on your dictionary. Here's an example that should remove Amy from our dictionary.

> 1. winCount.popitem()
> 2. print(winCount)

```
{'John': 40, 'Adam': 44, 'Michaela': 23, 'Mike': 4}
```

You may also use the **del keyword** we used before to remove a key-value or the entire dictionary itself. Here's how we remove John from our dictionary:

> 1. del winCount["John"]
> 2. print(winCount)

```
{'Adam': 44, 'Michaela': 23, 'Mike': 4}
```

Using the **clear** method, you can remove all key-value pairs from your dictionary and empty it, making it an empty dictionary:

1. winCount.clear()

2. print(winCount)

```
{}
```

There are several other functions and methods for making use of a dictionary but these essential ones should get you up and running in no time.

DAY 5

Decisions, Loops, Conditionals

5.1: Conditional Statements

5.1.1 If Statements

More often than not, you're faced with a situation where you have to decide something and then, a few things happen in response to your decision.

Similarly, programming languages also allow you to write conditional tests, with which, you can check a condition and make responses according to it. Let's take a real-life example, and then code it. **If** the stove is on, **then** close it. **If** it is not on, **then** do nothing.

If you take a look, we use the keyword 'if' when we're trying to put forth a condition. If this, then that. Likewise, Python uses the **if** statement to allow you to make a decision based on something. That 'something' in our example was, whether or not the stove was on. Let's code it.

1. # Whether or not the stove is on.
2. stoveOn = True
3.

```
4.  # Is the stove on?
5.  if stoveOn == True:
6.      print("The stove is on! Close it.")
```

Firstly, we declared our variable and said, yes, the stove is on. Now, using the **if statement,** we see, if the value of our variable when compared to the Boolean value, True, yields a True, i.e. the stove is actually on and both of their values are the same.

Now to the syntactical part. Using if statement is simple. You can either write your condition after the **if keyword** or you could wrap it in a pair of parentheses. Like this:

```
1.  if(stoveOn == True):
```

You can put in as many conditions as you want and use logical operators (and, or) to separate them and make a decision based on their values. Thirdly, after the condition, we use the ":" symbol to continue. Next, we tell what should our program do, if, the result of our conditional test is True. If not, it will completely ignore the code which follows this test. But, is it going to be the first following line only? Let's see.

Since in our case, it is True, we simply print a statement to close it right away. But, here's something odd. Do you see an indentation on the next line?

This indentation is how we recognize the part of the code which follows our if statement. In this case, there's only one statement. But, here's an example with multiple lines of code and quite a few conditions to test:

```
1.  # Whether or not the stove is on.
2.  stoveOn = True
3.  lightsOn = False
4.
5.  # Is the light on? Just check the stove.
6.  if(lightsOn == True):
7.    if(stoveOn == True):
8.      print("The lights are on, just close the stove.")
9.
10. # Is the light off? Open it, then, check the stove.
11. if(lightsOn == False):
12.   print("First, open the lights!")
13.   if(stoveOn == True):
14.     print("Close the stove!")
```

Now, we test a few conditions and then make our decision. We added another variable which helps us make a decision.

Firstly, check if the lights are on. If their values are correct, I indented the code in the next line, which means, lines 7-8 are to be executed only if the statement on line 6 is True.

If not, all that code is neglected. If, however, line 6 yields a True but line 7 returns a False. You see, I indented the code on line 8 again. Which means, this is now, a part of the if statement on line 7 and will not be executed in this case.

Now, on to the second conditional test. If the lights aren't on, print a statement to turn the lights on. Then check for the stoves. See, how the code on line 12-13 is indented once. It means they belong to the first if statement and will be executed by the interpreter if the statement on line 11 yields a True.

Similarly, code on line 14 only executes if the if statement on line 13 is True.

So, the **indentation in Python is important and must be well taken care of.** If you don't indent your code properly, either your conditions won't output a result, or, you will get an error.

5.1.2 'If-else' Statements

Now, what was our example again? Here's that sentence. **If** the stove is on, **then** close it. **If** it is not on, **then** do nothing.

You see, we catered the part where it says, if the stove is on, do this and that. But, what about the part where it isn't on? Let's take a look at this code. We expand on our first example.

```
1.  # Whether or not the stove is on.
2.  stoveOn = False
3.
```

> 4. # Is the stove on?
> 5. if stoveOn == True:
> 6.   print("The stove is on! Close it.")
> 7. else:
> 8.   print("The stove is off!")

Now, we put a False in our variable and continued our tests. If the stove is on (True output from the if condition), just print that it is on. If it is not, we use the **else** clause, to say, do this instead.

Like our real-life situations. Do this, else do that. So, what happens here? When the result from the if statement on line 5 is False, it neglects all indented code (which was to follow if that if statement was True) and executes the else clause.

Now, logically, it states, **If the stoveOn is False, do this.** And, we simply print, that the stove is already off!

```
The stove is off!
```

5.1.3 'if-elif-else' Statements

Usually, when asked to write multiple conditions, you might write them in a similar fashion: If this, then do this. Else if this, then do that. Else (none of these), do something completely different.

See, how all these if-elseif-else clauses are linked to one single conditional statement? This is where the **elif or Else If** block comes in.

If your variable, the one you wish to use for your conditional test, has many values, and needs to output differently for those values, you can put them in an if-elif-else block. This way, on the first true, no other condition gets executed. Or, elif gets executed, or the else clause.

We did just the same. We said, If the lights and stove are on, you just close the stove. **Else if (elif in Python),** the lights are closed but the stove is on, turn the lights, close the stove. And further, continue with the else statement.

```
1.  # Whether or not the stove is on.
2.  stoveOn = True
3.  lightsOn = False
4.
5.  # Is the light on? Just check the stove.
6.  if(lightsOn == True and stoveOn == True):
7.    print("The lights are on, just close the stove.")
8.
9.  elif (lightsOn == False and stoveOn == True):
10.   print("Turn the ligts on, close the stove!")
11.
12. else:
13.   print("Both the lights and the stove is on.")
```

```
Turn the ligts on, close the stove!
```

70

The example also shows how you can run multiple conditions in an if-elif-else statement and base the output on all those.

5.2: Loops

Loops are traditionally used whenever you have a block of code which you want to repeatedly use a couple of times. For that purpose, Python provides us the **for** statement and the **while** loop or statement. Let's see their differences and working next.

5.2.1: 'For' Loop

For loops can be used to iterate over the values or members of a list or any other sequence in a linear fashion, and executes the enclosed block each time.

Let's say, you wish to print the multiplication table of 5. Here's how you would do it using the for loop:

```
1.  number = 5
2.
3.  for i in range(1, 11):
4.      print(str(number) + " x " + str(i) + " = " + str(i*number))
```

If you execute the code, your output should be similar to this.

```
5 x 1 = 5
5 x 2 = 10
5 x 3 = 15
5 x 4 = 20
5 x 5 = 25
5 x 6 = 30
5 x 7 = 35
5 x 8 = 40
5 x 9 = 45
5 x 10 = 50
```

Now you know, that the for loop only executes for a fixed number of times. In our example, we use the for loop to iterate over the number 1 to 10 using the **range()** function.

The range function is used to provide a list of values that can be used for a variety of purposes. In our example, it produces a list of numbers from 1 to 10. The first argument is the starting number and the second argument is the ending number minus one. So, the 11 actually works till 10.

If you use range(7), it starts at 0 and makes a list of values from 0 to 6 with an increment of 1. If you wish to change the increment, provide that number as the third argument. So, range(1,10,2) will produce a list:

[1,3,5,7,9]

If the variable 'i' is lesser than 11, the block of code which is enclosed within the for loop keeps on executing. And in our example, it prints the table of 5 until 10.

Here's how you can iterate over a list using the for loop:

```
1. places = ['LA', 'SF', 'US', 'NV', 'TX']
2.
3. for place in places:
4.     print(place)
5.
```

When iterating over a list of elements, when we use the for loop, it uses a temporary variable (place in our example), which takes one value from the list (places in our example) and executes the code.

The **in** operator is used to check for membership. It basically checks for values in sequences like lists, tuples, or dicts, and proceeds. You may use an if statement to check if LA is in the list using this:

```
1. places = ['LA', 'SF', 'US', 'NV', 'TX']
2.
3. if('LA' in places):
4.     print("LA is in the list!")
```

And the output is: **LA is in the list!**

Back to our main example, on the first iteration, the value of place is **LA** which is printed, then SF, then US, then NV. Here's the output:

```
LA
SF
US
NV
TX
```

You can check for special conditions using the if statement as well:

```
1.  places = ['LA', 'SF', 'US', 'NV', 'TX']
2.
3.  for place in places:
4.      print(place)
5.      if(place == 'LA'):
6.          print(place + " is my favorite destination location!")
7.
```

Now, here's the output:

```
LA
LA is my favorite destination location!
SF
US
NV
TX
```

It checks each value of place (from the list places) and compares it to LA. If it comes up multiple times, the if clause will also be executed for the same number of times.

5.2.2: 'While' Loop

On the contrary, while loops are used to execute a block of code until a specific condition is me. As soon as that condition changes or just isn't True, the While loop breaks and the program terminates execution.

Here's a program to print a multiplication table of 5 using a while loop:

```
1.  number = 5
2.  maxNum = 1
```

```
3.
4.  while(maxNum != 10):
5.      print(str(number) + " x " + str(maxNum) + " = " + str(number*maxNum))
6.      maxNum += 1
```

Let's analyze the code. On the fourth line, we finally use our while loop and state that, until the maxNum variable doesn't hit the value 10, keep executing the block of code (indented block).

The '!=' symbol is used to denote inequality in Python. Next, in line 5, we do our printing by converting the integers to strings.

On line 6, we use the increment operator. This operator can also be written as **maxNum = maxNum + 1**. The shorter piece of code actually adds and assigns to the variable itself. Like other loops, all indented code is operated within the while loop itself.

Remember, if no condition is provided, the block of code runs forever (or something which always yields a True).

Here's an example program (your system might hang if you run this):

```
1.  number = 5
2.  maxNum = 1
3.
4.  while(maxNum != 10):
5.      print(str(number) + " x " + str(maxNum) + " = " + str(number*maxNum))
```

What's wrong here? **I removed the line where the maxNum variable was incrementing.** Now, its value is always 1 (non-zero is True), so, the loop never breaks.

5.3: Break and Continue

At times, you might want to break the loop out of its normal functioning. In this case, you may use the **break statement.**

In case, you wish to skip an iteration or if on some condition, you wish to neglect the logic inside the loop, you may use the **continue statement** and continue operations as if that value never existed.

Let's say we're looping over a string (using a for loop), but, if it contains a vowel, we want to break that loop. Here's how we do it:

```
1.  vowels = ['a', 'i', 'e', 'o', 'u']
2.  string = 'This is a string'
3.
4.  for letter in string:
5.      if letter in vowels:
6.          break
7.
8.      print("Letter: " + letter)
```

In this example, we make a list of vowels and take an exemplary string to loop over. In our loop, we state, if the letter variable is a vowel (if it exists in our list vowels using the **in membership operator)**, we break our loop. If not, we print the letter.

Now, let's see how we can make use of continue. We say, if the string contains a vowel, **continue**, otherwise, print the letter. Here's how we do it:

```
1. vowels = ['a', 'i', 'e', 'o', 'u']
2. string = 'Auspicious'
3.
4. for letter in string:
5.     if letter in vowels:
6.         continue
7.
8.     print("Letter: " + letter)
```

5.4: Exceptions

At times, we're faced with situations where errors might pop up but we're unsure if they will come up or not and don't wish to terminate the execution. Other than syntax errors, we have exception errors that come up on code which is logically correct Python code but still results in an error.

To cater such situations, Python has built-in exception handling techniques that will help us output the error or exceptions (in case something does go wrong, otherwise, the execution continues).

If you want your program to be flexible and have certain conditions it needs to meet, you can raise an exception error yourself using the **raise** command.

Here's how you do it:

```
1.  Numbers = [1, 3, 5, 8]
2.
3.  for num in Numbers:
4.      if num%2 == 0:
5.          #The number is even. The remainder will return a 0
6.          raise("An even number is found in an odd number list!")
```

We test to see, if any number in our list is an even number, we simply raise an error by using the raise command and provide the error we're going to display inside.

Here's the output to the program:

```
Traceback (most recent call last):
  File "main.py", line 23, in <module>
    raise("An even number is found in an odd number list!")
TypeError: exceptions must derive from BaseException
```

5.4.1: 'try...except' block

The try-except block is one of the most common exception handling technique in Python.

If you wish to run a risky block of code, in which you're unsure if it's going to work or not, you put that code in the **try** statement. This block of code is the **"normal"** execution of the code.

The code which you'll put in the except statement is the code which is the response you'll generate in case an exception is raised in the try clause.

In section 4.4, we saw that if an exception is unhandled, the program terminates its execution. Here's where, if we use the except statement, we can determine our response to the raised exception.

Here's an example for the try-except block:

1. print(5/0)

If you run this, you'll get a traceback error named **ZeroDivisionError**. This is a perfect example of the try-except block. If your user tries to run this faulty code, you can put this in the try block, and in the except block, simply tell the user of the mistake.

Here's the syntax of the try-except block:

1. try:
2. print(5/0) #This should raise an error because division by Zero isn't allowed
3. except:
4. print("Division by zero? Not allowed!")

Just like we said, we put the faulty or risky block of code in the try statement and to handle the exception error, we put it in the except block.

Here's the output:

```
Division by zero? Not allowed!
>
```

If you feel there are several errors which might come up, you can name the error in the except block like this:

1. try:

    print(x)

    except NameError:

    print("Define the variable 'x' and then use it. ")

    except:

      print("An undefined error came up. Look for it. ")

When we run this code, since, we've not defined the variable, we'll run the except block of **NameError.**

Here's the output:

```
Define the variable 'x' and then use it.
>
```

In case, we did define the variable x. What should happen?

1. x = "I'm a variable"
2. try:
3.   print(x)
4. except NameError:
5.   print("Define the variable 'x' and then use it. ")
6. except:

> 7. print("An undefined error came up. Look for it. ")

When we run this, since, there are no errors. The variable is printed as it is.

If you have a block of code which you wish to run, ONLY if there were no errors raised, you can append the **else:** block to the try-except statement. Here's how we use the else block:

> 1. x = "I'm a variable"
> 2. try:
> 3.   print(x)
> 4. except:
> 5.   print("An undefined error came up. Look for it. ")
> 6. else:
> 7.   print("No errors were raised in the execution.")

Now, the output is:

```
I'm a variable
No errors were raised in the execution.
>
```

If you have a block of code which you wish to run, DESPITE THE FACT that an error was raised or not, you can put it in the **finally:** block.

Here's how you use the finally block:

> 1. try:

2. print(x)
3. except NameError:
4. print("An undefined error came up. Look for it. ")
5. else:
6. print("No errors were raised in the execution.")
7. finally:
8. print("This ran although there was an exception error.")

The output of this program is:

```
An undefined error came up. Look for it.
This ran although there was an exception error.
>
```

Like we studied, the else block runs in case of no errors. Since there's an error here. Only the except block gets executed. Also, the finally block runs which were supposed to run with or without an error.

Finally is very helpful in running statements to save memory by closing files, or cleaning up resources after their usefulness has ended. (We'll go over this again when we discuss files).

If you want your program to ignore the raise of an exception error, you may use the **pass** command. Here's how you do it:

1. try:
2. print(5/0) #This should raise an error because division by Zero isn't allowed
3. except:

```
4.    pass
5.
6.    print("The error was safely ignored!")
```

PART 3

DIVING DEEPER INTO PYTHON

DAY 6

FUNCTIONS AND MODULES

6.1: What are Functions?

Functions allow us to write a block of code which we can re-use whenever we want in the code. Most functions perform operations which they are coded to do, but, functions can be dynamic as well (responsive to user inputs).

Wait, reuse code? Why don't you copy-paste it again?

Functions provide modularity to the application as it helps developers when the code is broken into smaller, more manageable sections. If you keep adding code on to code you're making it hard to debug in case of an error. For now, you have to update your changes everywhere you copied your faulty code. Functions help you reuse your code through a single, simple line again and again.

6.2: Writing Functions

Defining, writing, and using functions is fairly easy in Python. Let's try a different approach and write a sample function first, call it a few times, and discuss the output in relevance to the code. Open a file called basicFunctions.py, and save the following code:

```
1.  def myFunction(): #Define the function
```

```
2.    #All indented code is the body
3.    print("Just a plain ol' function!")
4.
5.
6.    #Call the function
7.    myFunction()
8.
9.    #Call it again
10.   myFunction()
```

Let's run it. Here's the output:

```
Just a plain ol' function!
Just a plain ol' function!
```

Now, what really happened here? In the first line, we use the **def** keyword or the statement to define our function. The def keyword is associated with a string. This string is the **function name.**

So, the statement becomes: **def functionName**

Next, the function name is ended with a pair of parentheses. These parentheses constitute the area where we use our parameters or arguments which are either sent to our function (sent from the calling statement) or made in the function itself (define ourselves).

Finally, these parentheses are ended with a colon and this marks the end of the def statement.

The final statement becomes: **def functionName():**

Next, all code which belongs to the function is indented so it's run with the function only. Whenever the function is executed by Python, these lines are executed as part of the function. Remember, these lines can't execute without **calling the function.**

This is what we do next, after removing the indentation, we begin to call the function. Our code 'myFunction()' is a function call. A simple call to a function is made by writing the function's name and appending the pair of parentheses to it. Like we discussed above, when we call the function, we can also pass some data we wish to use in the function.

So, finally, when we call the function twice, it runs twice, and the print statement (enclosed in the function) is executed twice.

Let's take a look at **parameters** next.

6.3: Parameter Values

From the start, we've been using the print() function and passing to it, whatever we wanted to print on the screen. These values we passed, are called **arguments**. Now, we can also make our own functions which can accept these arguments and use them in the body.

Here's an example program which makes use of arguments and parameters.

```
1.  def teamBuilder(team):
2.      print("Team: " + team)
3.
4.  teamBuilder('Raptors')
5.  teamBuilder('GSW')
```

And, the output of the program:

```
Team: Raptors
Team: GSW
```

Now, the working of the program:

In our definition statement, we have a **parameter** called *team*. So, what is a parameter? A parameter contains the value which was sent from the function call, as an argument.

All parameters (used in the parenthesis) are available for use in the function body. So, we make use of it in our print function as whenever we use 'team' it has the value which **was sent in that specific function call**. Once the program exits the function, all parameter values are forgotten. This is called scoping. Let's discuss this in a little more detail next.

A variable which is declared inside a function is only valid until the function is being evaluated and executed. As soon as it ends, these values will yield a **NameError** if called since they are undefined variables.

What if you pass two arguments to your function? It should work if and only if your function declaration allows two parameters to be taken. Here are a few errors you might come across. Try adding these lines and running them separately (to the same code we wrote before):

1. teamBuilder()

```
Team: Raptors
Team: GSW
Traceback (most recent call last):
  File "main.py", line 10, in <module>
    teamBuilder()
TypeError: teamBuilder() missing 1 required positional argument: 'team'
```

When you run this code, we get a TypeError since our function necessarily requires one parameter and we didn't provide that.

Next, let's pass more parameters than what the function can actually hold:

1. teamBuilder('Bucks', 'Suns')

Now, we get an error that our function only requires one argument and we're sending it two.

```
Team: Raptors
Team: GSW
Traceback (most recent call last):
  File "main.py", line 10, in <module>
    teamBuilder('Bucks', 'Suns')
TypeError: teamBuilder() takes 1 positional argument but 2 were given
```

Let's discuss the **return** keyword and how functions can terminate upon request as well.

6.4: Returning Values and the 'return' Statement

Before studying the return statement. Let's go over our print function once more. When you pass to it the string and run your program, what's the output? It's that same string displayed on the screen as part of your output.

In general, whenever a function yields an output after some evaluation or execution, it is called the **return value** of the function.

We can ask our functions to yield some return value using the **return keyword or statement.** So, a standard return statement includes the following two things:

1. The **return** keyword

2. The value(s) which are supposed to be returned from that function

Let's make a sample program where a decision is made and a relevant value is returned using the **return** statement. Here's a code:

```
1.  def findWinner(team, winRatio):
2.      if winRatio >= 0.90:
3.          return(team + ' has a winning chance!')
4.      elif winRatio >= 0.80 and winRatio <= 0.90:
5.          return(team + ' might win!')
6.      elif winRatio >= 0.70 and winRatio <= 0.80:
7.          return(team + ' has lesser chances!')
8.      else:
9.          return(team + ' has no chance!')
10.
11. decision = findWinner('GSW', 0.98a)
12. print(decision)
```

Here's the output to our program:

```
GSW has a winning chance!
```

Now, when we pass our arguments through the function call, they are firstly associated with the named parameters. Then, a conditional test series is run where we run some conditions on our winRatio variable and see if it matches a condition. Ultimately, we return a string statement through our function using the **return statement.**

Now, here's a fact. **The return statement causes the execution of the function to halt since it returns a value back to the calling function** and breaks the execution series. Now, you might've noticed, the return statement only passes the values and does nothing.

If you wish to use the returned values, you can use a variable and equate it with our function call. So, whenever the return statement returns something, it is caught by the variable and you may use the variable (which now has the returned values or the data).

If you don't want to store the data, simply put the function call in the print function. Here's how: (yields the same output as well)

> 1. print(findWinner('GSW', 0.98))

Now, although most functions you write will have a return statement. It is not necessary though. If your function returns nothing, it still returns a value which is called **None**. Remember, how your print function returns nothing other than outputting the string you sent to it? It returns a value of **None** as well. If no return statement is added to the end of a function, Python automatically adds a statement '**return None**' to the end of the function. If you simply write '**return**', then **None** is automatically appended to it.

6.5: More on Function Arguments

We saw a simple technique to pass data to our functions; through arguments. They were called **required arguments**; i.e. if they have to be in the function's declaration and are sent and received (at the function) in the same order.

Now, there are a few other types of arguments you may use to pass your data:

1. Keyword Arguments
2. Default Arguments
3. Variable-length Arguments

Keyword Arguments are those arguments where you assign your data to a variable and use the same variable (name) in the function's declaration. What's the advantage? You can now skip parameters since it is not ordered anymore and will only work if you pass the use the same name for your parameter.

Remember, **position arguments** (required arguments which have same order) **will always be declared/called before the keyword arguments.** Take a look at the code and the output:

```
1. def findWinner(winRatio, team):
2.     if winRatio > 0.9:
3.         return team
4.
5. decision = findWinner( 0.98, team = 'GSW')
6. print(decision)
```

```
GSW
```

Default arguments allow you to assume a value for your argument in the function's declaration. So, even if no such data is provided as an argument, it is still valid and present as a parameter. Here's a default parameter function in play:

```
1. def findWinner(team, winRatio = 0.5):
2.     if winRatio > 0.9:
3.         return team
4.
5. decision = findWinner(team = 'GSW', winRatio = 0.95)
6. print(decision)
7.
8. decision2= findWinner(team = 'Raps')
9. print(decision2)
```

```
GSW

None
```

Now, you may notice how in the first example, our winRatio variable goes as predicted. But, in the second function call, where we don't pass a ratio, it is assumed or defaulted to 0,5, the if condition is false, and nothing is returned; or None, so we studied.

Variable-length arguments allow you to bypass the argument limit in the function call. You may send one argument or you may send a hundred arguments. They will still be valid and treated appropriately. To do this, you have to place a * symbol before your last variable, which will act as the container for all arguments which follow your required or position arguments.

A tuple is formed with the data and can be accessed through a loop in the function. Here's how we use variable length arguments:

```
1.  def findWinner(*teams, winRatio = 0.5):
2.      if winRatio > 0.9:
3.          return teams
4.      else:
5.          return winRatio
6.
7.  decision = findWinner('Raptors', 'GSW', 'Bucks', winRatio = 0.95)
8.  print(decision)
9.
10. decision2= findWinner('Suns')
11. print(decision2)
```

```
('Raptors', 'GSW', 'Bucks')
0.5
```

Here, we pass 3 values in our argument list in the first call and only one in the second call. Just like we predicted, it works and prints all our arguments in the form of a tuple.

6.6: Scope of Variables

Variables which are defined in a program are not usable or accessible from every other line. It primarily depends on where the variable was first declared, and where is it going to be used.

It is the scope of the variable which helps us know where the variable can be used without throwing up errors. There are two scopes for variables:

- Global Scope
- Local Scope

Let's take a look at global scoping first. In the last few chapters, we simply declared our variables and used them wherever we wished for without a single error. It is because these variables have a **global scope** and are accessible in any function, any line of code, and any other loop as well.

Variables, which are sent through arguments (to become parameters) or declared inside the body of the function have a **local scope** and are only accessible inside the body of that specific function.

Let's take a look at an example first:

1. sum = 0 # Global variable

```
2.
3.  def addNums(num1, num2, num3 = 10):
4.      # Local variable 'sum'
5.      sum = num1 + num2 + num3
6.      print('Local SUM: ' + str(sum))
7.      return sum
8.
9.  print('Global SUM: ' + str(sum))
10.
11. finalSum = addNums(5,10)
12. print("Final Sum: " + str(finalSum))
```

```
Global SUM: 0
Local SUM: 25
Final Sum: 25
```

If you put our discussed concepts to use here, you'll know that on line 1, we declare our first global variable which can be used anywhere.

Next, we define a simple function which adds up 2-3 numbers. Inside the function, we use our global variable, sum. But, it becomes a local variable as it is used inside the function's body. Whatever you assign to sum, will only exist inside it until the function is existing or executed.

We tried this after our function. As soon as you print sum, you'll know, its value is still 0. If you were to make global changes to sum, they would show in your print statement. As an added exercise, try incrementing the sum (global var) in your code and print its updated value.

We call the function next and print the final sum, which is equal to **the local variable, sum, from the function's body.**

6.7: What are Modules?

The core purpose of functions was to increase modularity so debugging code and reusability was easier. To resolve this problem with larger code bases, **modules** were introduced.

Modules are smaller, logically relevant, and organized pieces of code which are kept separate in different files. It helps to understand the code when multiple people are working on a project.

A module can have everything from variables to functions, loops to classes (we'll discuss soon!) and other Python elements.

6.8: Writing Modules

Modules are nothing but simple Python files with logically grouped pieces of code. If you wish to use anything from a certain module, you can then import it, which we'll discuss in the next section.

But, for now, let's see what a simple module looks like and how to use it in some other file using an **import** statement:

myModule.py

```
1.  # Returns the sum of numbers
2.  def addNums(*numbers):
3.      sum = 0
4.      for number in numbers:
5.          sum += number
6.      return sum
```

mainFile.py

```
1.  import myModule
2.
3.  print(myModule.addNums(1,2,3,4,5))
```

In the file, myModule.py, we implemented a simple function which takes in a random number of variables and adds them up. Simple right? **This is a perfect example of a module. A file which has the code to add numbers separately and now, whenever you wish to add numbers, you can simply call this function in any call. How? Let's see!**

In our file, mainFile.py, the first line is an **import statement** which is used to bring the module file into the main file. We'll discuss the import statement in great detail next.

On our third line, we use the print function to print the output of the addNums function to the screen. What's important here is how it happens. Since the **addNums** function is a method of the **myModule** file, we use the dot (.) operator to call the function. The general syntax to call a function from a module is:

moduleName.modulesFunction or *moduleName.moduleVariable*

6.9: Importing Modules

Almost every python source file (the file which contains our source code) can be used as a module in another source file using the import statement.

The general syntax of an import statement is;

import module, module2, module3, ... module

As soon as this statement is interpreted by Python, it searches for the module in the basic search path; a directory where all modules are searched for when imported.

Now, the import statement can be changed however you like. If you wish to import a function from the module (only), you can import a single statement using the following syntax:

from module **import** functionName

In our example, it would be:

from myModule import addNums

Now, no other function from the module will be available. Also, now, you don't have to use the name of the module to call the function. Simply call it using the functions name: **addNums(...)**

Want to import all functions from a single module? You can do that too.

from module import *

The wildcard character, *, represents 'all' and can be used to import all functions into the current namespace – a dictionary of variable names (function names) and their values (bodies).

DAY 7

FILE MANAGEMENT WITH PYTHON

7.1: Handling Files

In real-life situations, you might be responsible for working with loads of data (stored in hundreds of files) and perform some analysis on it. The data is stored in files.

These files are either excel sheets, text files, or binary files (unusual, but exist). Let's discuss text files first. Most text files contain only one record (of data) per line and could be of any one of these schemes:

- CSV Files – Each record is separated by a single ','

  First Name, Middle Name, Last Name

  Adam, Jones, Marine

- Delimited Files – These files separate data values or records with a specific delimiter. It could be a '\t', '\n', or some symbol.

To work with files, we have to open them up first. This is possible using the **open() function**. We'll discuss it next, but, it returns your file as a special **Python Object** (called "handle"). We can use this object or handle to perform our operations.

As every opened file is treated as an object, they have their own attributes and methods which you can apply and manipulate your data however you want.

7.2: Opening Files in Python

To open a file in Python, we have a function called **open()** which takes in our file and a mode, and in return, provides us the python file object.

Here's the syntax for a general open function:

open(fileName, mode, buffering, encoding, errors, newline, closed, opener)

Too complicated, is it? Don't worry. For our initial operations, we only need the first two arguments. Most of them are given default values which help in performing smaller operations.

Now, let's code this. Firstly, make a text file in your directory titled 'textFile.txt' and then make a python file called 'fileHandling.py'. Here's the data you have to add in the text file, and the code in the python source file:

textFile.txt:

Just some random data

Add anything else you want to…

fileHandling.py:

```
1. myFile = open('textFile.txt')
2.
3. data = myFile.read()
4.
5. print(data)
```

Here's the output of the program:

```
Just some random data
Add anything else you want to..
```

We added some dummy data to our text file so we could use it in our Python source file. In our main file, we use the open function and pass to it the following arguments:

- fileName – This is the only required argument and doesn't have a default value. For the open function to run, you must have a valid file. You can either pass the name of the file (if it is in the same directory as your Python file) or pass the whole directory so it could access the file.

    Here are the two ways in code:

    ```
    1. myFile = open('textFile.txt')
    2. myFile = open('C:\\PythonFiles\\textFile.txt')
    ```

- Access mode – These modes define how and what operations will be valid for when you are successful in opening a file. These modes include open file for reading, writing, or for both.

The default value of the mode argument is **'r' or read**. So, if you try to write to it, you'll get an error (we'll discuss later).

Here is a table of modes you can use in your Python programs:

Symbol or Character	Mode Description
r	Open a file in read mode (default mode)
w	Open a file in write mode – Overwrites on existing data from the start - Creates a file if the named file doesn't exist
a	Open a file in append mode – Write data at the end of existing data – Creates a file if the named file doesn't exist
r+	Open a file for reading and writing both
w+	All features from 'w' mode and allows reading as well
a+	All features from 'a' mode and allows reading as well

Rb	Read a file in binary format
Wb	Write a file in binary format
Ab	Append in a file in binary format
wb+	All features from 'wb' mode and allows reading
ab+	Same features as 'ab' mode and allows reading as well

Since we provide no second argument, it is defaulted to 'r' or read mode. The rest of the arguments in the open function are also defaulted and we're not discussing them for now.

Now, to our 2nd line, the read function. Let's take a look at file operations next.

7.3: File Operations: Read, Write, Append

If you've copied and run the example, you know that we just tried our first file operation, Reading from a file. Let's go over the same example again, here's the code:

textFile.txt:

Just some random data

Add anything else you want to…

fileHandling.py:

```
1. myFile = open('textFile.txt')
2.
3. data = myFile.read()
4.
5. print(data)
```

Now, on line 3, we use the **read()** function on our file object (returned from the open function and contains our file) and then assign it to a variable. What goes in the variable? Take a look at the general syntax of the read function:

read(n) – where n is the number of bytes you wish to read from the file. If absent, the entire file is read.

Now, if you run the program, the entire file is outputted. Let's run it with a number, next. Replace line 3 with this:

```
1. data = myFile.read(10)
```

Print it again, this should be the output:

```
Just some
```

This shows that only 10 bytes were copied from the start (since the read mode opens the file and puts the pointer to the start).

If you don't want your program to read the file word by word, try the **readline() function** which reads a single line at once until a delimiter is approached. No single word is read after a line ends. Whenever a readline() function is observed and used, it replaces the current pointer and places it on the next line. So, if you run the readline() function again, it will yield the next line, and not the first line.

Let's try it out:

```
1. #Read the first line
2. print(myFile.readline())
3.
4. print("Second Line:")
5. print(myFile.readline())
```

```
Just some random data

Second Line:
Add anything else you want to..
```

Notice how after our first readline() method, there's a new line? It's the **End-of-Line Delimiter** which is copied from the file. So, the function automatically puts it to the new line.

You can also read a file using a for loop:

```
1. #Iterates over the buffer and prints
2. for aLine in myFile:
3.     print(aLine)
```

```
Just some random data

Add anything else you want to..
```

Enough reading these files. Let's get to writing already. It is just as simple, and we can do it using two main functions:

- write()

- writelines()

Let's work on a new example now. Take a look at the code below:

```
1.  fileObj = open('cities.txt', 'w')
2.
3.  fileObj.write('London\n')
4.  fileObj.write('Bombay\n')
5.  fileObj.write('Shanghai\n')
```

'w' stands for writing so we opened a file in write mode and assigned it to a handle. Again, the write mode creates the required file if it doesn't exist and overwrites on the existing data if it does. So, in our case, it makes the file and begins operations.

We use the **write()** function and pass to it, the string, we wish to write in our file. See, the delimiters have to be put in there by yourself so they can be used a new line in the outputted file. Let's take a look in our directory (provided or the one where we are right now) and see if there's a file.

```
cities.txt        saved
1    London
2    Shanghai
3    Bombay
4
```

The file is created and data is added to it. The python code worked just as predicted. **REMEMBER** to close your file before running your program to see the results in action. Otherwise, most of these changes are stuck in the buffer and do not actually show.

Next is the **writelines()** function which takes a list as input and writes all the elements in the file. Here's an example:

```
1.  cities = ['Karachi\n', 'Kabul\n']
2.
3.  fileObj = open('cities.txt', 'a+')
4.
5.  fileObj.writelines(cities)
6.
7.  fileObj.close()
```

If you go see the file, cities.txt, now, you will see the list is now added:

```
cities.txt      saved  ▼
1    London
2    Shanghai
3    Bombay
4    Karachi
5    Kabul
6    Karachi
7    Kabul
8
```

7.4: A Brief Discussion on Binary Files

Binary files are usually composed of several different sequences of bytes, 1s, and 0s. These files are most commonly used to indicate images or exe files which have lesser text representation and more numeric, byte data.

Working on binary files is similar to text files. All you have to do is use the appropriate modes like 'wb', 'rb', and 'ab', and there should be no issues. However, unlike a text file, binary files can't be opened easily so only store digital data there and not text data you wish to see for yourself and not ask a computer program to resolve for you.

7.5: Closing Files

Once we're done with the processing, it's better to close the file using the Python way to ensure all changes are made successfully and no data loss has occurred.

You can do this using the **close()** method on the file handler (the object you receive from the opening of file). It will make sure the buffer is empty and close the file, relieving the system of all taken memory.

Here's the close function in play:

```
1.   myFile.close()
```

If you want to get rid of closing files yourself, here's a newer method of opening files which involves auto-closing of files. The **with** keyword or statement allows you to open a file and handle the file object just like your open function, but automatically flushes the buffer and saves the changes.

Here's an example:

```
1.   with open('cities.txt', 'r') as fileObj:
2.       print(fileObj.read())
```

```
London
Shanghai
Bombay
Karachi
Kabul
Karachi
Kabul
```

Now, if you try to close the file, you'll get an error since the file object doesn't exist anymore and can't be closed. As an added exercise, try the examples we wrote above using the **with** statement.

DAY 8

Object-Oriented Programming

8.1: Introduction to Object-Oriented Programming

Object-oriented programming (OOP) is a programming paradigm in which programs are modeled according to their properties and behaviors rather than functions and logic. All these elements are then bundled into objects.

Let's say, for example, an object could be you or me in real life. A person, with a valid name, age, birth date, occupation, and other data, or **properties** in terms of programming languages. Also, we have certain **behaviors**. We can walk, talk, work, sleep, jog, and others as well.

So, OOP allows us to program and model real-world elements and make them as realistic and meaningful as possible. Each entity in the world can be modeled as a Python object which possesses some data and does some function (has some behavior).

What have we been doing till now? It's the procedural programming paradigm. It provides steps, functions, and code blocks which follow a sequential order of completing commands.

Let's take a look at the most basic concepts of OOP; **Classes**.

8.2: Classes and Objects

To model real-world objects in programming, we need a blueprint of these objects or a prototype on which these objects will be based on. **Classes** are basically user-defined blueprints which state how an object should look, what attributes or properties its object should have, and what should it do (the behaviors).

Basically, we describe the general behavior each object of a class can have. What are objects? Objects are *instances* of a class, what we work with in life and programs. This process, making objects from classes, is called instantiation.

Let's take an example under consideration.

If you've ever come across a car, let's see what attributes it can have. Color, number of tires, a model of the car, engine specifications, and others. When we program our class called **Car,** these will act as the properties of our car.

Now, what does the car do? It drives, honks, and performs other functions internally. These are the **properties or methods** of our class **Car.** See, how every car performs these actions and has these properties; **Classes are general representations of real-world objects.**

Objects, are the specific instances of these classes and have relevant data in them. For example, a Ford Mustang will be different from an SUV and have massively different properties. Both of them are individual objects from our class Car.

8.3: Writing Classes

Let's head back to our editor and code an example class with properties and behaviors. Here's the code: (don't stress, I'll explain everything later)

```
1.  class Car:
2.      '''Modelling a car'''
3.
4.      def __init__(self, model, license):
5.          '''Initialize all attributes and properties '''
6.          self.model = model
7.          self.license = license
8.
9.      def drive(self):
10.         print("Vroom vroom! The car drives!")
11.
12.     def honk(self):
13.         print("HONK! HONK!")
1.  fordMustang = Car("ford-8", "AX-2939")
2.  SUV = Car("Honda", "MX-2101")
```

Now, on to the analysis of the code we just wrote.

8.4: An Explanation on Classes (Code Breakdown)

We begin by defining our class on line one using the **class** keyword and immediately following it is the name of the class. Conventionally, we start the name of the class in uppercase letters.

On line 2, we define a **docstring**. It is a simple statement which tells us more about what the class has to offer or what it does.

On line 4, we finally define a function, since we know that functions are defined using the def keyword. All functions defined in a class are called **methods** of that class. The **__init__** is a special method provided by Python for every class, which, upon the instantiation process, runs automatically (when you create a new object).

A question you might have: Why the underscores? They are to help you understand that it is a default function provided by Python and it shouldn't conflict with your own special function names.

Now, it takes in three parameters in our case, but it can have as many parameters as you want. The **self** parameter is necessary and should come before others. What is self? **The self keyword is a reference which helps objects refer to themselves anywhere in the class. It allows objects to have individual access to all the properties and methods defined in the class and doesn't interfere with other objects.**

The self keyword is automatically passed whenever an object is made and all other parameters can be passed with it (optional, but if used in the class declaration, they must be provided).

Now, on line 6, we prefix each parameter with self. This is so, each object of the class has its own attributes (specific to it) and can be used throughout the class for that object only.

Next, we define two other functions and pass the self parameter to it, which is necessary so each object has access to its own methods.

This is it for our class, let's see what happens next.

8.5: Making an Instance: Objects

Out of the scope of the class, we are finally using our class to make objects of it (or cars out of the class Car). These are basically instructions for how our class should behave for a specific car.

We can make an object using this syntax:

```
1.  objectOfTheClass = nameOfClass('param1', 'param2', …)
```

Let's see how we did it for our example:

```
1.  fordMustang = Car("ford-8", "AX-2939")
2.
3.  SUV = Car("Honda", "MX-2101")
```

Simply, we ask Python to make a car whose model is something and the license is something else. Again, we ask Python to make a different car with different data.

How does it work? As soon as you instantiate an object and assign it, the interpreter runs the __init__ function and assigns self to the newly made object and also associates the passed arguments to the parameters. The init method then returns an object and it is assigned to our variable fordMustang.

Now, let's use this object to see what attributes or properties our objects have.

8.6: Accessing Attributes and Methods

Try running the following code after instantiating your class:

```
1. print(fordMustang.model)
2. print(fordMustang.license)
```

It prints what we sent to it using the arguments in our class. As they are associated with our object now, the self.model is used to send back the data to us. Here's the output:

```
ford-8
AX-2939
```

If you ask for these attributes from the second object, the output will be what you sent with it. Here's an example:

```
1. print(SUV.model)
2. print(SUV.license)
```

```
Honda
MX-2101
```

Now, if you want to access the methods, simply use the dot operator again and ask for the methods. Here's how:

```
1. print(fordMustang.honk())
```

And it outputs:

```
HONK! HONK!
None
```

'None' is actually the return statement which is executing and printing as well.

Let's write a new method and use an attribute to see different outputs for different objects: (Add to your class from the last example)

```
1.    def mileage(self):
2.        val = input("What is the mileage? ")
3.        print(self.model + " Mileage: " + val)
```

When you run this on a class, you will be prompted to enter a value since we use the **input() function.** Enter the value and let's check the output:

print(fordMustang.mileage())

```
What is the mileage? 2200
ford-8 Mileage: 2200
None
```

Similarly, you can run this in the second class.

Let's take a look at some other concepts for Object-oriented programming, next.

This chapter covers a little more advanced topics from object-oriented programming like inheritance, child classes, and others. Also, we'll see how to import classes just like we imported modules.

8.7: Inheritance

In real-world situations, most objects have a relationship to other objects. Similarly, if we program something which is a specialized version of a more general element, this programming concept is called **Inheritance** where a child class grabs all properties and methods from the parent class and makes use of them and adds something of its own.

The parent class is the class which is more general and has all the basic functions. For example, if we wrote the code for a Car, it is pretty general. If now, we wish to write a class for an electric car, it will inherit most of the properties and behaviors from the parent (Car) class and add more stuff of its own.

Let's take a look at child classes next.

8.8: Child Classes: Writing One

Let's add to our code base from the last chapter by making a child class of our class, Car. We'll model an electric car, a more specific form of our Car class.

Here's the code and let's analyze it afterwards:

```
1.  class Car:
2.
3.      def __init__(self, model, license):
4.          self.model = model
5.          self.license = license
6.
7.      def drive(self):
8.          print("Vroom vroom! The car drives!")
9.
10.     def mileage(self):
11.         val = input("What is the mileage? ")
12.         print(self.model + " Mileage: " + val)
13.
14. class ElectricCar(Car):
15.
```

```
16.    def __init__(self, model, license):
17.        super().__init__(model, license)
18.
19. teslaX = ElectricCar("Tesla", "AA-9323")
20.
21. print(teslaX.mileage())
22. print(teslaX.model)
```

```
What is the mileage? 343
Tesla Mileage: 343
None
Tesla
```

Firstly, we write our child class and use the parenthesis to provide to it, the parent class (Car).

Next, we declare the __init__ function just like we did before and pass to it the parameters and the self keyword to refer to the object.

Next, something strange. We use the **super()** function and use the method __init__ to refer to the **init method of the parent class. This is done so a connection can be made between the parent class and the child class and now, it can access all attributes and methods of the parent class.**

Although it doesn't have any function of its own right now, it can definitely be added in later.

Next, we make an object of our new ElectricCar class and ask for the methods and attributes which yield expected output since now, a relationship is made between the **parent and child or super and subclass.**

If you decide to assign methods to the child class, remember, the parent class can't access them. But, the child class can definitely (always) access the methods of the parent class.

8.9: Importing Classes

As your programs grow, so will the complexity, both in logic and file size. It is always recommended to ship your classes as individual files and import them wherever they are required. This is possible using the import statements we studied a while ago.

Here's how you can import the classes into another file and use them properly:

1. from car import Car
2. from car import ElectricCar
3. from car import Car, ElectricCar
4. from car import *
5. import car

Made in the USA
Columbia, SC
30 July 2019